KT-529-186

The Oxford Book of Animal Poems

Oxford University Press, Walton Street, Oxford OX2 6DP

Oxford   New York   Toronto
Delhi   Bombay   Calcutta   Madras   Karachi
Petaling Jaya   Singapore   Hong Kong   Tokyo
Nairobi   Dar es Salaam   Cape Town
Melbourne   Auckland

and associated companies in
Berlin   Ibadan

Oxford is a trade mark of Oxford University Press

This selection and arrangement ©
Michael Harrison and Christopher Stuart-Clark 1992

First published 1992
First published by Oxford in the United States 1992

All rights reserved. No part of this publication may be reproduced, stored in a retrieval
system, or transmitted, in any form or by any means, without the prior permission in
writing of Oxford University Press.
Within the U.K., exceptions are allowed in respect of any fair dealing for the purpose
of research or private study, or criticism or review, as permitted under the Copyright,
Designs and Patents Act, 1988, or in the case of reprographic reproduction in
accordance with the terms of the licences issued by the Copyright Licensing Agency.
Enquiries concerning reproduction outside those terms and in other countries should
be sent to the Rights Department, Oxford University Press, at the address above.

Library of Congress Catalog Card Number: 91-051150
A CIP catalogue record for this book is available from the British Library

ISBN 0 19 276105 6

Typeset by Tradespools Ltd., Frome, Somerset
Printed in Hong Kong

This book is to be returned on or before
the last date stamped below.

3 0 0 MAR 1993 PROJECTS

2 4 2 JUN 1993

3 16 JUN 1994

22 9 MAR 1995

29 5 JUN 1995

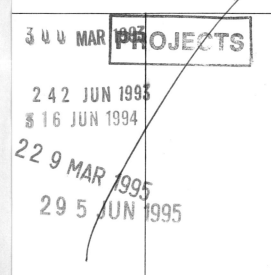

SCHOOL LIBRARY SERVICE
CENTRAL REGIONAL COUNCIL

Hillpark Education Centre,
Benview,
BANNOCKBURN,
Stirling.
FK7 OJY

821
OXF

Copy 001

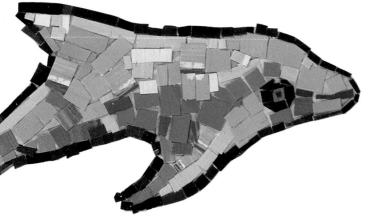

The Oxford Book of Animal Poems

Michael Harrison
and Christopher Stuart-Clark

Oxford University Press

Oxford   New York   Toronto

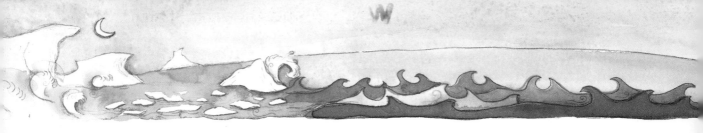

Contents

Copy No.   001

Class No.   821.008

Author   OXF

Gondwanaland

It is thought that millions of years ago, South America, Africa, Arabia, India, Australia and Antarctica were all one huge mass of land—known now as Gondwana. When they drifted apart, the continents each took with them their own animals.

O take me back to Gondwanaland
Before the Terrible Split
When what is now cloven
Was all interwoven
And didn't differ a bit!

O dinosaurs of Gondwanaland,
Now so vanished away,
I miss you sorely,
I ache so rawly,
Won't you come back one day?

Marsupials of Gondwanaland,
Bigger than elephants too,
Sloths and possums,
Odontoglossums,
Why do I dream of you?

Steamy heat of Gondwanaland,
When reptiles were the kings!
Life's so mammalian
And neo-Australian
And lots of *un*pleasant things!

O waft me back to Gondwanaland,
Fly me away in the sky—
For ever I'm all
In thrall to what's primal,
Until the day that I die!

Gavin Ewart

Octopus

Marvel at the
Awful many-armed
Sea-god Octopus,
And the coiled
Elbows of his eager
Eightfold embrace;

Yet also at his
Tapered tender
Fingertips, ferrying
Their great brow
Along the sea floor
In solitary grace.

Valerie Worth

Skate

Flitting the sea-bed, wide and flat,
I am a fish to wonder at:

a kind-of, sort-of soft thick square,
like a slow plane in watery air,

with a tough white spine from tail to lips
and straight bones out to my side-wing-tips.

No other interests interest me,
being this is my whole activity.

Alan Brownjohn

The Names of the Sea-trout

He who would seek her in the clear stream,
Let him go softly, as in a dream,
He who would hold her well,
Let him first whisper the spell
Of her names.

The silver one, the shimmering maiden,
The milkwhite-throated bride,
The treasure-bringer from the sea,
Leaper of weirs, hurdler to the hills,
The returning native, egg-carrier,
The buxom lass, the wary one,
The filly that shies from a moving shadow,
The darter-away, the restless shiner,
Lurker in alder roots,
The fearful maid.

Night dancer, ring maker,
The one that splinters reflections,
The splasher, the jester, the teaser, the mocker,
The false encourager, tweaker of lures,
The girl who is fasting, destroyer of hopes,
Bender of steel, the breaker, the smasher,
The strong wench, the cartwheeler,
The curve of the world,
She who doesn't want to surrender,
The desired, the sweet one.

When you've spent nights and days
Speaking her names, learning her ways,
Take down your tackle from the shelf,
And your skill. She may give herself
For the whispered spell.

Tom Rawling

The Shark

He seemed to know the harbour,
So leisurely he swam;
His fin,
Like a piece of sheet-iron,
Three-cornered,
And with knife-edge,
Stirred not a bubble
As it moved
With its base-line on the water.

His body was tubular
And tapered
And smoke-blue,
And as he passed the wharf
He turned,
And snapped at a flat-fish
That was dead and floating.
And I saw the flash of a white throat,
And a double row of white teeth,
And eyes of metallic grey,
Hard and narrow and slit.

Then out of the harbour,
With that three-cornered fin,
Shearing without a bubble the water
Lithely,
Leisurely,
He swam—
That strange fish,
Tubular, tapered, smoke-blue,
Part vulture, part wolf,
Part neither—for his blood was cold.

E. J. Pratt

Sea Lions off Monterey

Half a mile into the Pacific
a rock like the back of a sea-beast
humps itself out; the swell sounds
hollowly as it bumps the sides even
in slack water, the sides the only bits
visible under the sea lions
crowding the top, still, mostly,
as clusters of mussels.
                     On shore
round the coin-op telescope
the children bounce and wrangle,
and the cars slide down to the shore
and slide away again. An encounter
on the edge of things, the sea's curled
lip, the end of a continent,
the abounding other, leaving
nothing to be said.
                   The still,
head-up animals stare
indifferently out of the ocean.

John Cassidy

Seal Lullaby

Oh! hush thee, my baby, the night is behind us,
And black are the waters that sparkled so green.
The moon, o'er the combers, looks downward to find us
At rest in the hollows that rustle between.
Where billow meets billow, there soft be thy pillow;
Ah, weary wee flipperling, curl at thy ease!
The storm shall not wake thee, nor sharks overtake thee,
Asleep in the arms of the slow-swinging seas.

Rudyard Kipling

Dolphins

The dolphins play in the sea like children,
Diving, leaping ahead of the long ship,
Rising out of the sparkling waves;
    Their jaws smile,
    'Friends', they seem to say,
    'We are friends, play with us.'

The water sprays from their curving backs
And in the bubbles of their shining wake.
    Small fish watch the game
    Round-eyed and
    Open-mouthed.

However fast the ship, the dolphins are faster,
Sliding joyfully through the emerald sea.
At last descending into the deepest water,
They swim away from the ship and the bright sky.

Zoë Bailey

Dolphins

for Tom Durham

They've brains the size of a man's and they like music—
surely make it . . . May be exchanging speech?

        We (it seems) catch their low
notes only—
        and they, of human music,
        may read more than we know—
firm shapes, shot with elusive depths, of dapples
subtly disturbed by thrusts of arabesque,
        the way a floor mosaic
        mixes its own clear message
in with the high riches, the dappled panels.

Do dolphins stop just short of words? talk music?
Having preferred purity, think in music?

Jonathan Griffin

The Song of the Whale

Heaving mountain in the sea,
Whale, I heard you
Grieving.

Great whale, crying for your life,
Crying for your kind, I knew
How we would use
Your dying:

Lipstick for our painted faces,
Polish for our shoes.

Tumbling mountain in the sea,
Whale, I heard you
Calling.

Bird-high notes, keening,
Soaring:
At their edge a tiny drum
Like a heartbeat.

We would make you
Dumb.

In the forest of the sea,
Whale, I heard you
Singing.

Singing to your kind.
We'll never let you be.
Instead of life we choose

Lipstick for our painted faces,
Polish for our shoes.

Kit Wright

24

Whalesong

I am
ocean voyager,
sky-leaper,
maker of waves;
I harm no man.

I know
only the slow tune
of turning tide,
the heave and sigh
of full seas meeting land
at dusk and dawn,
the sad whale song.
I harm no man.

Judith Nicholls

Penguin

*'The male penguin . . . picks up a pebble in his bill, waddles
over to a bird standing alone and solemnly lays it before it
. . . if the stranger receives the pebble with a deep bow then
he has discovered his true mate.'*
David Attenborough: *Life on Earth*

This stone I set at your feet
As my courtship gift to you
At the white summer's end
On Antarctica's icy shore.

Later you lay your egg
And ease it on to my feet.
You turn and walk away,
Black going into the blackness.

I stand, Emperor of this land,
My back to the blistering wind,
Shifting my feet with care
Through the dark of sixty days.

Will you return as the egg hatches,
Fat, and with belly full to feed
Our young? Will you find us here
Amidst this blinding snow?

Yes, and I must now walk
My shrunken self the hundred icy
Miles to open sea, then, fattened,
The hundred back, my belly lardered.

And so through all the long dark
We trudge, or stand, under the howling sky.
In the fading summer I will bring again
The honest gift of a stone.

Michael Richards

Sea-hawk

The six-foot nest of the sea-hawk,
Almost inaccessible,
Surveys from the headland the lonely, the violent waters.

I have driven him off,
Somewhat foolhardily,
And look into the fierce eye of the offspring.

It is an eye of fire,
An eye of icy crystal,
A threat of ancient purity,

Power of an immense reserve,
An agate-well of purpose,
Life before man, and maybe after.

How many centuries of sight
In this piercing, inhuman perfection
Stretch the gaze off the rocky promontory,

To make the mind exult
At the eye of a sea-hawk,
A blaze of grandeur, permanence of the impersonal.

Richard Eberhart

From 'Australia'

All day the kookaburra is laughing
from the phantoms of trees,
from the satire of nature.
It is not tragedy nor comedy,
it is the echo of beasts,
the bitter chorus of thorns,
and flowers that have names
that aren't easily remembered.
The kookaburra laughs from the trees,
from the branches of ghosts,
but the sky remains blue
and the eyes glow green in the night.

Iain Crichton Smith

Egrets

Once as I travelled through a quiet evening,
I saw a pool, jet-black and mirror-still.
Beyond, the slender paperbarks stood crowding;
each on its own white image looked its fill,
and nothing moved but thirty egrets wading—
thirty egrets in a quiet evening.

Once in a lifetime, lovely past believing,
your lucky eyes may light on such a pool.
As though for many years I had been waiting,
I watched in silence, till my heart was full
of clear dark water, and white trees unmoving,
and, whiter yet, those thirty egrets wading.

Judith Wright

30

Galahs

There are about fifty of them
on the stony ground,

some standing still,
some moving about.

Nothing much of pink
breast or lighter-hued crest

shows in the twilight
among the stones.

They are standing about
like little grey-coated aldermen

talking in undertones.

William Hart-Smith

Ringneck Parrots

The ringneck parrots, in scattered flocks,
The ringneck parrots are screaming in their upward flight.

The ringneck parrots are a cloud of wings;
The shell parrots are a cloud of wings.

Let the shell parrots come down to rest,
Let them come down to rest on the ground!

Let the caps fly off the scented blossoms!
Let the blooms descend to the ground in a shower!

The clustering bloodwood blooms are falling down,
The clustering bloodwood blossoms, nipped by birds.

The clustering bloodwood blooms are falling down,
The clustering bloodwood blossoms, one by one.

Aranda, Australia
Translated by T. G. H. Strehlow

Parrots

parrots
with vermilion bands and beak
green-iris camouflaging
are acrobats
swinging on trapezes of green gum leaves
tips

they carry their very own safety net
their green-yellow tail feathers
which spray out like palm fronds
parachuting

Neil Paech

Old Man Platypus

Far from the trouble and toil of town,
Where the reed-beds sweep and shiver,
Look at a fragment of velvet brown—
Old Man Platypus drifting down,
Drifting along the river.

And he plays and dives in the river bends
In a style that is most elusive;
With few relations and fewer friends,
For Old Man Platypus descends
From a family most exclusive.

He shares his burrow beneath the bank
With his wife and his son and daughter
At the roots of the reeds and the grasses rank;
And the bubbles show where our hero sank
To its entrance under water.

Safe in their burrow below the falls
They live in a world of wonder,
Where no one visits and no one calls,
They sleep like little brown billiard balls
With their beaks tucked neatly under.

And he talks in a deep unfriendly growl
As he goes on his journey lonely;
For he's no relation to fish nor fowl,
Nor to bird nor beast, nor to horned owl;
In fact, he's the one and only!

A. B. Paterson

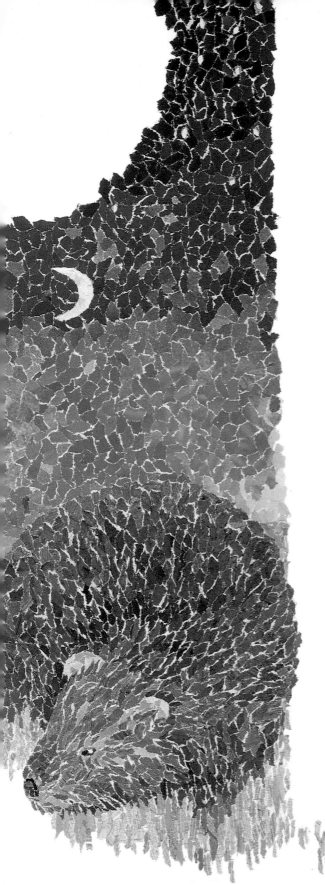

Weary Will

The strongest creature for his size
But least equipped for combat
That dwells beneath Australian skies
Is Weary Will the Wombat.

He digs his homestead underground,
He's neither shrewd nor clever;
For kangaroos can leap and bound
But wombats dig for ever.

The boundary-rider's netting fence
Excites his irritation;
It is to his untutored sense
His pet abomination.

And when to pass it he desires,
Upon his task he'll centre
And dig a hole beneath the wires
Through which the dingoes enter.

And when to block the hole they strain
With logs and stones and rubble,
Bill Wombat digs it out again
Without the slightest trouble.

The boundary-rider bows to fate,
Admits he's made a blunder,
And rigs a little swinging gate
To let Bill Wombat under.

So most contentedly he goes
Between his haunt and burrow:
He does the only thing he knows,
And does it very thorough.

A. B. Paterson

35

Under the Range

Where the gully shadows lie
Deeply blue before the sun,
In the shadow of the range
Wallaby are on the run.

Where the cliffs reach to the sky
In sudden mountains darkly strange,
Where the pine and mallee dare
To climb the jagged range.

He makes his home that none will find,
The wallaby, the secret one,
Alone he runs his soakage pad,
Alone he sits when day is done.

Sits like a statue on the rocks,
His little striped face bright and wise,
And nothing stirs there that is not
Reflected in his eyes.

Is he lonely there at night
When the quiet of night is on?
Does he think the light blown out
If suddenly the stars are gone?

Does he mind the bitter winds
When the evening light has fled?
Does he find a cave of rocks
With a dry earth bed?

Out there the fires of stars are set,
Now frosty winter has begun,
To keep a watch at night for those
Rock wallaby upon the run.

Irene Gough

From 'Kangaroo'

Delicate mother Kangaroo
Sitting up there rabbit-wise, but huge, plumb-weighted,
And lifting her beautiful slender face, oh! so much more gently
    and finely lined than a rabbit's, or than a hare's,
Lifting her face to nibble at a round white peppermint drop
    which she loves, sensitive mother Kangaroo.

Her sensitive, long, pure-bred face.
Her full antipodal eyes, so dark,
So big and quiet and remote having watched so many empty
    dawns in silent Australia.

Her little loose hands, and drooping Victorian shoulders.
And then her great weight below the waist, her vast pale belly
With a thin young yellow little paw hanging out, and straggle
    of a long thin ear, like ribbon,
Like a funny trimming to the middle of her belly, thin little
    dangle of an immature paw, and one thin ear.

D. H. Lawrence

Buffalo Country

Out where the grey streams glide,
Sullen and deep and slow,
And the alligators slide
From the mud to the depths below
Or drift on the stream like a floating death,
Where the fever comes on the South wind's breath,
There is the buffalo.

Out on the big lagoons,
Where the Regia lilies float,
And the Nankin heron croons
With a deep ill-omened note,
In the ooze and the mud of the swamps below
Lazily wallows the buffalo,
Buried to nose and throat.

From the hunter's gun he hides
In the jungles dark and damp,
Where the slinking dingo glides
And the flying foxes camp;
Hanging like myriad fiends in line
Where the trailing creepers twist and twine
And the sun is a sluggish lamp.

On the edge of the rolling plains
Where the coarse cane grasses swell,
Lush with the tropic rains
In the noon-tide's drowsy spell,
Slowly the buffalo grazes through
Where the brolgas dance, and the jabiru
Stands like a sentinel.

All that the world can know
Of the wild and the weird is here,
Where the black men come and go
With their boomerang and spear,
And the wild duck darken the evening sky
As they fly to their nests in the reedbeds high
When the tropic night is near.

A. B. Paterson

If You Go Softly

If you go softly out to the gum trees
At night, after the darkness falls,
If you go softly and call—
    Tch, Tch, Tch,
    Tch, Tch, Tch,
    They'll come—
    the possums!

If you take bread that you've saved
They'll come close up, and stand
And eat right from your hand—
    Softly,
    Snatching,
    Nervous—
    the possums!

And if you are still, and move slowly,
You can, very softly, pat
Their thick fur, gently, like that—
    It's true!
    You can!
    Really touch them—
    the possums!

You can do that all—
If you go softly,
At night,
To the gum trees,
If you go softly
—and call.

Jenifer Kelly

40

Possums

We've possums in our roof—how very sweet!
You'd think I'd hear the patter of their feet.
You'd think I'd wake sometimes from peaceful sleep
Aroused by gentle rustling as they creep
On rafters in our spider-muffled loft.
You'd think I'd hear them scamper, velvet-soft,
These smoky shadows flitting overhead
With delicate and dainty-tripping tread.

Huh!

They thunder round the racetrack of the beams,
Then organize themselves in football teams;
Their games are much like ours are, on the whole—
I'll swear I've heard triumphant yells of 'Goal!',
A frightful thud as two of them collide,
And uproar as they bellow out 'Offside!'
Then scuffles, whacks and wallops as they fight—
A thumping possum rumpus in the night.

Ann Coleridge

Night Herons

It was after a day's rain:
the street facing the west
was lit with growing yellow;
the black road gleamed.

First one child looked and saw
and told another.
Face after face, the windows
flowered with eyes.

It was like a long fuse lighted,
the news travelling.
No one called out loudly;
everyone said 'Hush'.

The light deepened; the wet road
answered in daffodil colours,
and down its centre
walked the two tall herons.

Stranger than wild birds, even,
what happened on those faces:
suddenly believing in something,
they smiled and opened.

Children thought of fountains,
circuses, swans feeding:
women remembered words
spoken when they were young.

Everyone said 'Hush';
no one spoke loudly;
but suddenly the herons
rose and were gone. The light faded.

Judith Wright

Flying Foxes

They drift down the dusk
Like sticks on a river,
Now in twos and in threes,
Now a dozen together.

Quiet as shadows
Their current is flowing,
Where from and where bound
Far outside of my knowing,

Till, passing my plum-tree,
First one, then another
Are clinging, like debris
Thrown clear by the water.

And then ah, the clatter!
The frenzy of eating!
The wings of black leather
All flapping and beating!

Their day is the night;
Their night is the day;
And long before sunrise
They're prisoned away.

Nothing is left
But a leaf-littered lawn,
And fruit on the garden-bed,
Tooth-marked and torn.

Lydia Pender

Orang-utan

Watch me,
touch me,
catch-me-if-you-can!
I am
soundless,
swung-from-your-sight,
gone with the wind,
shiver of air,
trick-of-the-light.

Watch me,
touch me,
catch-me-if-you-dare!
I hide, I glide,
I stride through air,
shatter the day-star dappled light
over forest floor.
The world's in my grasp!
I am windsong,
sky-flier,
man-of-the-woods,
the arm of the law.

Judith Nicholls

Ape

I am the Ape and I can climb
    straight up the tallest wall,
up flagpoles, maypoles and North Poles,
    yes, anything at all
that's vertical, horizontal or
    even diagonal.
I never hesitate because
    if I did I would fall.
I swing from branch to branch of trees
    so sickeningly tall
that I could look down far upon
    St Paul's Cathedral
and spit a little bit of spittle
    upon the Dome of Paul.
Yes, far below me I survey
    a world so very small
that I could take it in my hand
    like a little rubber ball—
O I must be the Principal
    and Primate of it all.
I am the Ape, the highest and
    the cleverest animal.

George Barker

Autumn Cove

At Autumn Cove, so many white monkeys,
bounding, leaping up like snowflakes in flight!
They coax and pull their young ones down from the branches
to drink and frolic with the water-borne moon.

Li Po
Translated by Burton Watson

The Legend of the Panda

Many years ago we were all white;
But then a leopard stole one of our young.
A brave young girl ran quickly up to help,
She saved the cub but, wounded by the leopard,
Died. Then we all sadly came together
To show respect to this young rescuer;
And as a sign of grief we pure white bears
Wore shawls of black. But as we rubbed our tears
The shawls left stains of black around our eyes;
And as we held our arms around our heads
In grief, the shawls soon stained our noses black;
And as the black shawls touched our ears they too
Turned black. So now we are both white and black
In memory of that brave young heroine.

Anthony Stuart

Bei-shung

I am *Bei-shung*, they call me the white bear.
I am the hidden king of these bamboo forests,
Invisible with my white fur and my black fur
Among this snow, these dark rocks and shadows.

I am the hidden king of these mountain heights,
Not a clown, not a toy. I do not care
To be seen. I walk, for all my weight,
Like a ghost on the soles of my black feet.

Invisible with my black fur and my white fur
I haunt the streams. I flip out little fishes;
I scoop them out of the water with my hand.
(I have a thumb, like you. I have a hand.)

Among this sparkling snow, these rocks and shadows,
I roam. Time is my own. My teeth are massive.
My jaw is a powerful grinder. I feed
On chewy bamboo, on small creatures, fish, birds.

You call me Panda. I am King *Bei-shung*.

Gerard Benson

The Tiger

Tiger! Tiger! burning bright
In the forests of the night,
What immortal hand or eye
Could frame thy fearful symmetry?

In what distant deeps or skies
Burnt the fire of thine eyes?
On what wings dare he aspire?
What the hand dare seize the fire?

And what shoulder, and what art
Could twist the sinews of thy heart?
And, when thy heart began to beat,
What dread hand? and what dread feet?

What the hammer? what the chain?
In what furnace was thy brain?
What the anvil? what dread grasp
Dare its deadly terrors clasp?

When the stars threw down their spears,
And water'd heaven with their tears,
Did he smile his work to see?
Did he who made the lamb make thee?

Tiger! Tiger! burning bright
In the forests of the night,
What immortal hand or eye
Dare frame thy fearful symmetry?

William Blake

50

Ballad of the Ferocious Tiger

Where did this tiger come from?
He has invaded the village of Three Forks!
Yesterday he devoured the eastern neighbour's pig,
today he has eaten the western neighbour's dog!
The animals of the mountain dare not make a sound;
dark winds blow in the poplars and bamboo.
A youth from one family, feeling very brave,
goes off into the mountains, with his bow and arrow.
Late at night, beneath a black moon, he stalks the tiger . . .
and the next day, only his white bones are found,
                    in the wild grass.

Hsu Pen
Translated by Jonathan Chaves

The Butterfly

There is no story behind it.
It is split like a second.
It hinges around itself.

It has no future.
It is pinned down to no past.
It's a pun on the present.

It's a yellow butterfly.
It has taken these wretched hills
under its wings.

Just a pinch of yellow,
it opens before it closes
and closes before it o

where is it

Jejuri Arun Kolatkar

In the Garden

Two ink-blue butterflies,
Their probosces curled,
Slumber on a sunflower
Like moored yachts
In a quiet harbour
With gaudy sails unfurled.

A great bumble-bee goes humming by
Like a Zeppelin in the sky,
Encasing the night in each eye.

Rupendra Guha Majumdar

The Shot Deer

Tear on tear
Weeps the dew,
As the grey deer
Slips to view.
And her spouse
Cries to her:
From shadows
Of deep boughs.
Do not stir!
Questing food,
The swift moving—
Through wide wood,
The light-loving—
Fears no check.
Till a hum—
In her meek
Slender neck
Shivers, dumb!
Fleetness, vain!
And love, too!
Red flowers stain
Deer and dew.

Hindi, India

India

They hunt, the velvet tigers in the jungle,
The spotted jungle full of shapeless patches—
Sometimes they're leaves, sometimes they're hanging flowers,
Sometimes they're hot gold patches of the sun:
They hunt, the velvet tigers in the jungle!

What do they hunt by glimmering pools of water,
By the round silver Moon, the pool of Heaven:
In the striped grass, amid the barkless trees—
The Stars scattered like eyes of beasts above them!

What do they hunt, their hot breath scorching insects,
Insects that blunder blindly in the way,
Vividly fluttering—they also are hunting,
Are glittering with a tiny ecstasy!

The grass is flaming and the trees are growing,
The very mud is gurgling in the pools,
Green toads are watching, crimson parrots flying,
Two pairs of eyes meet one another glowing—
They hunt, the velvet tigers in the jungle.

W. J. Turner

A Glimpse of Shere Khan

So the jungle
was pard and barred,
reflexive with shadows
patterning our path
as the coir saddle rocked slowly.
We were all eyes.

Then Bashir, mahout,
straddling behind the beast's ears
gripped my wrist hard.
'Tiger,' his teeth hissed
as our elephant began
a long, strong bout
of shuddering through and through,
head on to where
stripery had been and gone.

Around, back and up right,
canting to cut him off
in some douce clearing
and—
        HIST!—
                so we did:
all aglow now, his yellow
astripe, approaching, grand,
he had us heart in mouth.
We stiffened. He stood,
shade on his hide,
the whole dry glade
gone tense. It seemed an age
then, flick, he was lost,
filtered through thin jungle
slotted in space-time.

Chris Wallace-Crabbe

The Law of the Jungle

Now this is the Law of the Jungle—as old and as true as the sky;
And the Wolf that shall keep it may prosper, but the Wolf that shall break it
  must die.

As the creeper that girdles the tree-trunk the Law runneth forward and back—
For the strength of the Pack is the Wolf, and the strength of the Wolf is the Pack.

Wash daily from nose-tip to tail-tip; drink deeply, but never too deep;
And remember the night is for hunting, and forget not the day is for sleep.

The Jackal may follow the Tiger, but, Cub, when thy whiskers are grown,
Remember the Wolf is a hunter—go forth and get food of thine own.

Keep peace with the Lords of the Jungle—the Tiger, the Panther, the Bear;
And trouble not Hathi the Silent, and mock not the Boar in his lair.

When Pack meets with Pack in the Jungle, and neither will go from the trail,
Lie down till the leaders have spoken—it may be fair words shall prevail.

When ye fight with a Wolf of the Pack, ye must fight him alone and afar,
Lest others take part in the quarrel, and the Pack be diminished by war.

The Lair of the Wolf is his refuge, and where he has made him his home,
Not even the Head Wolf may enter, not even the Council may come.

The Lair of the Wolf is his refuge, but where he has digged it too plain,
The Council shall send him a message, and so he shall change it again.

If ye kill before midnight, be silent, and wake not the woods with your bay,
Lest ye frighten the deer from the crops, and the brothers go empty away.

Ye may kill for yourselves, and your mates, and your cubs as they need, and ye can;
But kill not for pleasure of killing, and *seven times never kill Man*.

If ye plunder his Kill from a weaker, devour not all in thy pride;
Pack-Right is the right of the meanest; so leave him the head and the hide.

The Kill of the Pack is the meat of the Pack. Ye must eat where it lies;
And no one may carry away of that meat to his lair, or he dies.

The Kill of the Wolf is the meat of the Wolf. He may do what he will,
But, till he has given permission, the Pack may not eat of that Kill.

Cub-Right is the right of the Yearling. From all of his Pack he may claim
Full-gorge when the killer has eaten; and none may refuse him the same.

Lair-Right is the right of the Mother. From all of her year she may claim
One haunch of each kill for her litter, and none may deny her the same.

Cave-Right is the right of the Father—to hunt by himself for his own:
He is freed of all calls to the Pack; he is judged by the Council alone.

Because of his age and his cunning, because of his gripe and his paw,
In all that the Law leaveth open, the word of the Head Wolf is Law.

Now these are the Laws of the Jungle, and many and mighty are they;
But the head and the hoof of the Law and the haunch and the hump is—Obey!

Rudyard Kipling

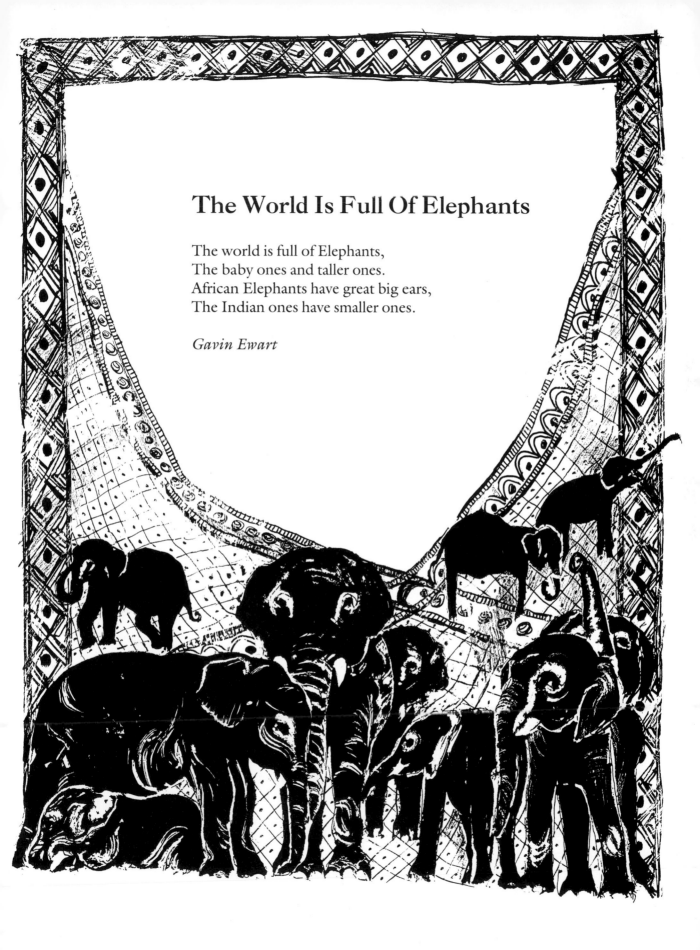

The World Is Full Of Elephants

The world is full of Elephants,
The baby ones and taller ones.
African Elephants have great big ears,
The Indian ones have smaller ones.

Gavin Ewart

Elephant

It is quite unfair to be
obliged to be so large, so I suppose
you could call me discontented.

Think big, they said, when
I was a little elephant; they
wanted to get me used to it.

It was kind. But it doesn't help if,
inside, you are carefree in small ways,
fond of little amusements.

You are smaller than me, think
how conveniently near the flowers are,
how you can pat the cat by just

halfbending over. You can also
arrange teacups for dolls, play
marbles in the proper season.

I would give anything to be
able to do a tiny, airy, flitting
dance to show how very little a

thing happiness can be really.

Alan Brownjohn

The Meerkats of Africa

Meerkats go about in packs,
They don't hang loose—
They're not really *cats* at all,
But more a mongoose.
They have great capabilities,
Make no mistake,
A Meerkat can kill a scorpion
Or even a snake.
They rescue each other's children
And have lookouts when they're feeding
And a system of babysitters—
The kind of co-operation
That the human race is needing!

Gavin Ewart

Chain-Song

If a jackal bothers you, show it a hyena,
If a hyena bothers you, show it a lion,
If a lion bothers you, show it an elephant,
If an elephant bothers you, show it a hunter,
If a hunter bothers you, show him a snake,
If a snake bothers you, show it a stick,
If a stick bothers you, show it a fire,
If a fire bothers you, show it a river,
If a river bothers you, show it the wind,
If the wind bothers you, show it God.

Fulani, West Africa

The Locust

What is a locust?
Its head, a grain of corn; its neck, the hinge of a knife;
Its horns, a bit of thread; its chest is smooth and burnished;
Its body is like a knife-handle;
Its hock, a saw; its spittle, ink;
Its underwings, clothing for the dead.
On the ground—it is laying eggs;
In flight—it is like the clouds.
Approaching the ground, it is rain glittering in the sun;
Lighting on a plant, it becomes a pair of scissors;
Walking, it becomes a razor;
Desolation walks with it.

Madagascan
Translated by A. Marre and Willard R. Trask

Scorpions Fighting

They run and throb a black tattoo
On the rawhide drum of earth;
Dark, crouched and fateful
Under the blaring daylight.

They shoot their racing shadows and themselves
Across the surface of their battleground,
In dancing lethal pageantry.
Their shadows race again,
The to, fro flicker of their pincers
Clasping mightily,
And up above their scalding tails thrust, poised,
In fiery armoury.

Black, hateful little flags,
Under the sun's dark patronage.

Then a great baboon hand drops down from the sky,
Grasping one combatant, whips off the hellish sting
And thrusts the active body into yellow teeth,
Still writhing.

The scorpion surviving,
From the rocks stands shining,
Triumphant and afraid.

Broughton Gingell

The Snake Song

Neither legs nor arms have I
But I crawl on my belly
And I have
Venom, venom, venom!

Neither horns nor hoofs have I
But I spit with my tongue
And I have
Venom, venom, venom!

Neither bows nor guns have I
But I flash fast with my tongue
and I have
Venom, venom, venom!

Neither radar nor missiles have I
But I stare with my eyes
And I have
Venom, venom, venom!

I master every movement
For I jump, run and swim
And I spit
Venom, venom, venom!

John Mbiti

The Crocodile

The crocodile is full of spleen:

He mulls the bile of judges.

He purifies his river;
the white corpuscle
devours the red.

His law
is the law of take
and it is just.

The water
thrashes
at his moment of truth.

Time and he continue.

Bruce Hewett

The Camel

Lord,
do not be displeased.
There *is* something to be said for pride
against thirst, mirages
and sandstorms;
and I must say
that, to face and rise above
these arid desert dramas,
two humps
are not too many,
nor an arrogant lip.
Some people criticise
my four flat feet,
the base of my pile of joints,
but what should I do
with high heels
crossing so much country,
such shifting dreams,
while upholding my dignity?
My heart wrung
by the cries of jackals and hyenas,
by the burning silence,
the magnitude of Your cold stars,
I give You thanks, Lord,
for this my realm,
wide as my longings
and the passage of my steps.
Carrying my royalty
in the aristocratic curve of my neck
from oasis to oasis,
one day shall I find again
the caravan of the magi?
And the gates of Your paradise?

                                    Amen.

Carmen Bernos de Gasztold
Translated by Rumer Godden

Ostrich

Large cold farms in sandy places
—to be an ostrich is serious and difficult.

An eagle has a dignity already, an ostrich
has to try to make his own.

Ostriches have to learn to be thought stupid
and not mind it,
have to learn to stand about in gawky, plumed groups
and be laughed at by a lot of foolish people
as a lot of foolish birds
and not resent it.

Little ostriches
fumble out of their big eggs to be told,
'Patience is the thing, dear, an impatient ostrich
is making trouble for himself.'

An ostrich's life is a hard life which you sometimes
feel you could run away from.
Sometimes you even think it just won't
bear looking at any more.

But in the end it's a worthwhile
    job with good prospects and if you
        still want to apply to be an ostrich
            send for the forms today.

Alan Brownjohn

Mother Parrot's Advice to her Children

Never get up till the sun gets up,
Or the mists will give you a cold,
And a parrot whose lungs have once been touched
Will never live to be old.

Never eat plums that are not quite ripe,
For perhaps they will give you a pain;
And never dispute what the hornbill says,
Or you'll never dispute again.

Never despise the power of speech;
Learn every word as it comes,
For this is the pride of the parrot race,
That it speaks in a thousand tongues.

Never stay up when the sun goes down,
But sleep in your own home bed,
And if you've been good, as a parrot should,
You will dream that your tail is red.

Ganda, Africa
Translated by A. K. Nyabongo

The Red-winged Lourie

Slowly he spread his wings
Like a fan,
A splash of red on the sky's blue page.
Enchanted, I watched him.
Slowly he sank,
Deep in the arms of a giant tree.
I waited.
I watched him feed
On fig-like fruit.

'Surely,' I thought,
'He'll soon sing.'
And he did,
But his grace was no paean of beauty,
Just a shock.

A most hideous noise came from his throat.
Slowly, raucously:
'Nobody's perfect,' he croaked.

Noëline Barry

The Gazelle

Fleet,
light,
my heart stifled
with wild fear,
always ready to leap
away with the wind,
at the least noise,
the least cry,
I bless You
Lord,
because for me
You have set no bounds
to Your space;
and if I fly,
an arrow
on my slender legs,
my little hooves
barely skimming the ground,
it is not that I scorn
the peace of Your pools,
but so, Lord,
that my life
might be a race
run straight
to the haven of Your love.

                                    Amen.

Carmen Bernos de Gasztold
Translated by Rumer Godden

Kob Antelope

A creature to pet and spoil
An animal with a smooth neck.
You live in the bush without getting lean.
You are plump like a newly-wedded wife.
You have more brass rings round your neck
　　than any woman
When you run you spread fine dust
Like a butterfly shaking its wings.
You are beautiful like carved wood.
Your eyes are gentle like a dove's.
Your neck seems long, long
　　to the covetous eyes of the hunter.

Yoruba, Nigeria
Translated by Ulli Beier

Zebra

White men in Africa,
Puffing at their pipes,
Think the zebra's a white horse
With black stripes.

Black men in Africa,
With pipes of different types,
Know the zebra's a black horse
With white stripes.

Gavin Ewart

The Giraffes

I think before they saw me the giraffes
Were watching me. Over the golden grass,
The bush and ragged open tree of thorn,
From a grotesque height, under their lightish horns,
Their eyes were fixed on mine as I approached them.
The hills behind descended steeply: iron-
Coloured outcroppings of rock half covered by
Dull green and sepia vegetation, dry
And sunlit: and above, the piercing blue
Where clouds like islands lay or like swans flew.

Seen from those hills the scrubby plain is like
A large-scale map whose features have a look
Half menacing, half familiar, and across
Its brightness arms of shadow ceaselessly
Revolve. Like small forked twigs or insects move
Giraffes, upon the great map where they live.

When I went nearer, their long bovine tails
Flicked loosely, and deliberately they turned,
An undulation of dappled grey and brown,
And stood in profile with those curious planes
Of neck and sloping haunches. Just as when,
Quite motionless, they watched I never thought
Them moved by fear, a wish to be a tree,
So as they put more ground between us I
Saw evidence that these were animals
With no desire for intercourse, or no
Capacity.
              Above the falling sun,
Like visible winds the clouds are streaked and spun,
And cold and dark now bring the image of
Those creatures walking without pain or love.

Roy Fuller

Vulture

On ragged black sails
he soars hovering over
everything and death;
a blight in the eye
of the stunning sun.

An acquisitive droop
of beak, head and neck
dangles, dully angling,
a sentient pendulum
next to his keeled chest.

His eyes peer, piously
bloodless and hooded,
far-sighted, blighting
grasses, trees, hill-passes,
stones, streams, bones—ah, bones—

with the tacky slack
of flesh adherent.
A slow ritual fold
of candid devil's palms
in blasphemous prayer—

the still wings sweep closed—
the hyena of skies
plummets from the pulpit
of a tall boredom,
swallowing as he falls.

He brakes lazily
before his back breaks
to settle on two
creaky final wing-beats
flinging twin dust-winds.

He squats once fearfully.
Flushed with unhealthy plush
and pregustatory
satisfaction, head back,
he jumps lumpishly up.

Slack neck with the pecked
skin thinly shaking, he
sidles aside then stumps
his deliberate banker's
gait to the stinking meal.

Douglas Livingstone

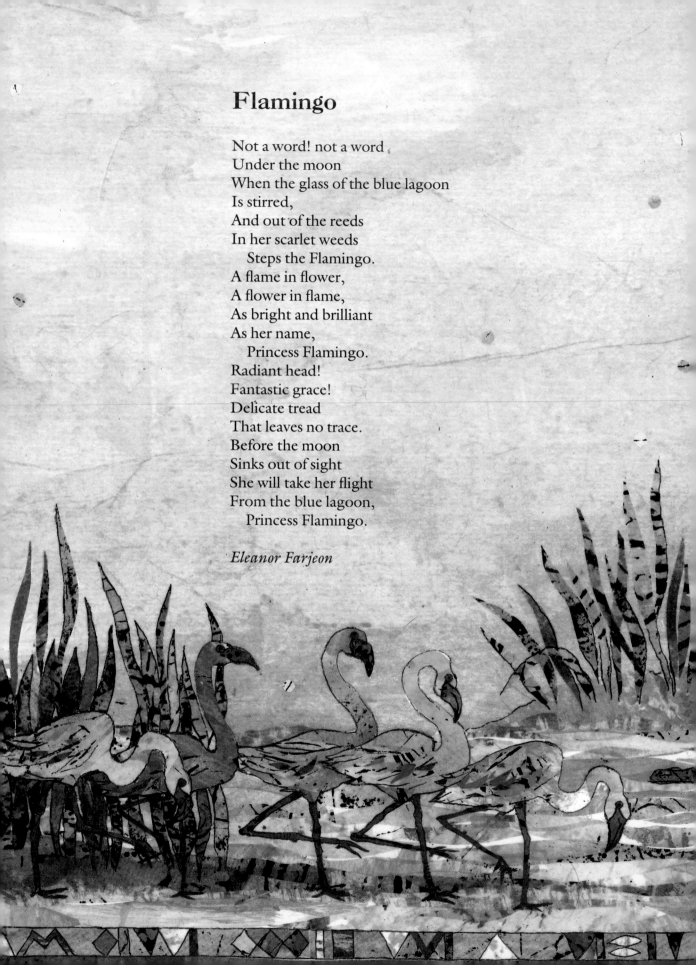

Flamingo

Not a word! not a word,
Under the moon
When the glass of the blue lagoon
Is stirred,
And out of the reeds
In her scarlet weeds
    Steps the Flamingo.
A flame in flower,
A flower in flame,
As bright and brilliant
As her name,
    Princess Flamingo.
Radiant head!
Fantastic grace!
Delicate tread
That leaves no trace.
Before the moon
Sinks out of sight
She will take her flight
From the blue lagoon,
    Princess Flamingo.

Eleanor Farjeon

Hippo

The hippo is a visual joke,
inflated like a rubber bed,
whose little bulging features poke
out of a yawning head.

Her jaws are built to match her girth,
with palate ribbed for emphasis.
Full-fed, she flops upon the earth
and smiles with rosy bliss.

But when she launches from the verge
to swim with curved aquatic grace,
those nostrils, eyes and ears emerge
above a hidden face—

and as by underwater bulk
you're comprehensively surveyed . . .
is hippo an ungainly hulk?
Or marvellously made?

Margaret Toms

The Praises of Baboon, the Totem Animal

Masters of the forests.
Those who make the rain to fall.
And who make all kinds of caterpillars come.
Sons of the Leader of big groups.
Their young clinging on to their backs.
Those of the long faces.
Smooth as a log without any bark.
Watchmen.
Those of the gruff voices.
Swaying to and fro in the trees;
Swarming here and there in the fields.
In the rains and just before always searching for food.
Animal almost human.
Knowing where the wild plums and the wild loquats are.

'I climb Rupara with my teeth;
And on the summit I cry defiance.
But among ants I hold Tom Thumb with my foot.'

Fellows with unkempt hair,
Their body knows water only when it rains.
One who takes everything,
Even the wild bulbs and wild onions here.

'I forage all over, a bull of Tingini.
Walking as if I do not see,
But I see very well indeed.
Among scorpions as well, I hold Tom Thumb in my foot.
Walking, you would say I was planting ground-peas.
And running, you'd think I was planting ground-nuts.'

Shona, Zimbabwe
Translated by George Fortune

The Theology of Bongwi, the Baboon

This is the wisdom of the Ape
   Who yelps beneath the Moon—
'Tis God who made me in His shape
   He is a Great Baboon.
'Tis He who tilts the moon askew
   And fans the forest trees,
The heavens which are broad and blue
   Provide him his trapeze;
He swings with tail divinely bent
   Around those azure bars
And munches to his Soul's content
   The kernels of the stars;
And when I die, His loving care
   Will raise me from the sod
To learn the perfect Mischief there,
   The Nimbleness of God.

Roy Campbell

Buffalo

The buffalo is the death
that makes a child climb a thorn tree.
When the buffalo dies in the forest
the head of the household is hiding in the roof.
When the hunter meets the buffalo
he promises never to hunt again.
He will cry out: 'I only borrowed the gun!
I only look after it for my friend!'
Little he cares about your hunting medicines:
he carries two knives on his head,
little he cares about your danegun,
he wears the thickest skin.
He is the butterfly of the savannah:
he flies along without touching the grass.
When you hear thunder without rain—
it is the buffalo approaching.

Yoruba, Nigeria
Translated by Ulli Beier

Rhinoceros

Ants and birds trace patterns in the dirt, but these creatures,
Armageddon in their shoulders, slip out of sight,
Sun at the meridian, and we are afraid to move
During the interregnum of the afternoon
Lest we encounter their colossal shadows,
Centres of gravity that flatten the grass
And range with unimaginable violence
Over the countryside we have rashly entered.

It is a landscape from which men are absent,
But not because they have migrated to the towns.
The stumps and trunks of trees clutter our path, and in the odd clearing
Evidence remains of human habitation,
Thatching grass and roof-poles not altogether destroyed by fire.
In these villages they paused and then disappeared,
Swept up like shadows in the aftermath of the sun.

We do not mind where we go, provided we do not meet them,
The missing people who occupied this savannah.
Trespassing in their pillaged territory
We might find the rhinoceros, we might hear him
Stamping the earth to tears.

Harold Farmer

Hyena

I am waiting for you.
I have been travelling all morning through the bush and not eaten.
I am lying at the edge of the bush
on a dusty path that leads from the burnt-out kraal.
I am panting, it is midday, I found no water-hole.
I am very fierce without food and although my eyes
are screwed to slits against the sun
you must believe I am prepared to spring.

What do you think of me?
I have a rough coat like Africa.
I am crafty with dark spots
like the bush-tufted plains of Africa.
I sprawl as a shaggy bundle of gathered energy
like Africa sprawling in its waters.
I trot, I lope, I slaver, I am a ranger.
I hunch my shoulders. I eat the dead.

Do you like my song?
When the moon pours hard and cold on the veldt
I sing, and I am the slave of darkness.
Over the stone walls and the mud walls and the ruined places
and the owls, the moonlight falls.
I sniff a broken drum. I bristle. My pelt is silver.
I howl my song to the moon—up it goes.
Would you meet me there in the waste places?

It is said I am a good match
for a dead lion. I put my muzzle
at his golden flanks, and tear. He
is my golden supper, but my tastes are easy.
I have a crowd of fangs, and I use them.
Oh and my tongue—do you like me
when it comes lolling out of my jaw
very long, and I am laughing?
I am not laughing.
But I am not snarling either, only
panting in the sun, showing you
what I grip
carrion with.

I am waiting
for the foot to slide,
for the heart to seize,
for the leaping sinews to go slack,
for the fight to the death to be fought to the death,
for a glazing eye and the rumour of blood.
I am crouching in my dry shadows
till you are ready for me.
My place is to pick you clean
and leave your bones to the wind.

Edwin Morgan

Hyena

The scruffy one
who eats the meat
together with the bag
in which it is kept.
The greedy one
who eats the mother
and does not spare the child.
God's bandy-legged creature.
Killer in the night.

Yoruba, Nigeria
Translated by Ulli Beier

On the Serengeti

A yellow lioness
is like a flame
to a herd of impala.

Ears wide
eyes huge with fear
and tails flicking in alarm,
they jump every way
like fire crackers

and the lioness
hesitating, for one brief second,
between two fireworks
finds the air clear

until one springs
out forwards, for her pride,
waiting.

Marilyn Watts

Herd of Impala

If I close my eyes I can hear them:
A herd of impala, leaping
Across a clearing level as a beach
Strewn with burnt mopani branches turning yellow.

I would go a long way to hear that sound,
A whoosh, a whisper like a thin sword from its scabbard,
A sound like the gleam of a bayonet on night guard,
The far-off rustle of bushes being brushed on soft ground.

They break like pieces from a noiseless grenade,
Death-dealing armour, bullets fired from a rubber gun,
Arcs from each day's birthday of the sun,
Or calling lovers, the final hiss of the last train.

Their horns arc lancers on patrol, pennantless thorns in winter,
The shaking of nude pines in torches of frost,
Brown waves, forever duplicated trough on trough,
Dropping trophies on the beaches, butt ends from the ocean and fins.

Yet, not so beautiful; unlike hunters I cannot discriminate
The off-cuts for the hunting lodge, the plungers for museums.
I am a marksman only in values of humanness—
The wires upon their bones, the scars of traps, the final helplessness as bait.

Colin Style

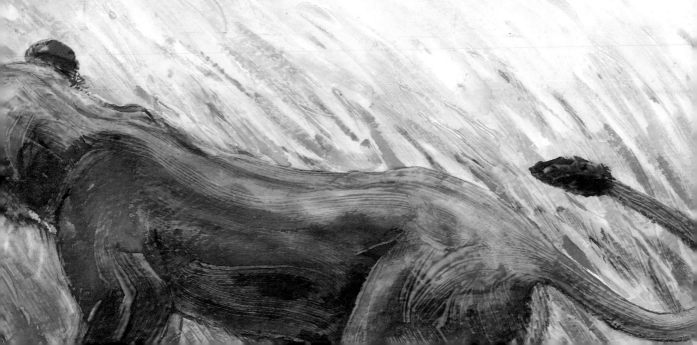

Song of the Lioness for her Cub

Fear the one
who has sharp weapons
who wears a tassel of leopard tail,
he who has white dogs—
O son of the short-haired lioness!
My short-eared child,
Son of the lioness who devours raw flesh,
you flesh-eater!
Son of the lioness whose nostrils are red with the bleeding prey
you with the bloodred nostrils!
Son of the lioness who drinks water from the swamp,
You water-drinker!

Khoikhoi, South Africa
Translated by Ulli Beier

Cheetah

Indolent and kitten-eyed,
This is the bushveld's innocent—
The stealthy leopard parodied
With grinning, gangling pup-content.

Slouching through the tawny grass
Or loose-limbed lolling in the shade,
Purring for the sun to pass
And build a twilight barricade

Around the vast arena where,
In scattered herds, his grazing prey
Do not suspect in what wild fear
They'll join with him in fatal play;

Till hunger draws slack sinews tight
And vibrant as a hunter's bow:
Then, like a fleck of mottled light,
He slides across the still plateau.

A tremor rakes the herds: they scent
The pungent breeze of his advance;
Heads rear and jerk in vigilant
Compliance with the game of chance

In which, of thousands, only one
Is centred in the cheetah's eye;
They wheel and then stampede, for none
Knows which it is that has to die.

His stealth and swiftness fling a noose
And as his loping strides begin
To blur with speed, he ropes the loose
Buck on the red horizon in.

Charles Eglington

Leopard

See the golden Leopard with the spots!
The golden cat of the cliffs!
See the Leopard with the bulging cheeks,
The golden Leopard with the wide face, I-Fear-Nothing,
The particoloured one, I-Climb-Into-A-Small-Tree,
I rip off the eyebrows!
Clawer am I, dig my nails in deep,
My enemies I leave behind, saying
'This was not one leopard but ten!'
Mr Claws, Scratch-For-Yourself,
Even a strong man is not ashamed to howl when clawed!
Leopards of the Tlokwa country,
Wild cat with the wide face,
We eat the wild antelope and the tame cattle.
The great golden spotted one,
Lone outlaw who brings thousands to him by his art,
Whose victim goes off with his scalp over his eyes,
Leopard of many spots,
Dark-spotted Leopard,
Fierce old man Leopard,
Even when his teeth are gone, he kills his prey with his head!

Sotho, Africa

You!

You!
Your head is like a hollow drum.
You!
Your eyes are like balls of flame.
You!
Your ears are like fans for blowing fire.
You!
Your nostril is like a mouse's hole.
You!
Your mouth is like a lump of mud.
You!
Your hands are like drum-sticks.
You!
Your belly is like a pot of bad water.
You!
Your legs are like wooden posts.
You!
Your backside is like a mountain-top.

Igbo, Nigeria

The Legend

Some say it was seven tons of meat in a thick black hide
you could build a boat from, stayed close to the river
on the flipside of the sun where the giant forests were.

Had shy, old eyes. You'd need both those hands for *one*.
Maybe. Walked in placid herds under a jungly, sweating roof
just breathing, a dry electric wind you could hear a mile off.

Huge feet. Some say if it rained you could fish in a footprint,
fruit fell when it passed. It moved, food happened, simple.
You think of a warm, inky cave and you got its mouth all right.

You dream up a yard of sandpaper, damp, you're talking tongue.
Eat? Its own weight in a week. And water. Some say
the sweat steamed from its back in small grey clouds.

But *big*. Enormous. Spine like the mast on a galleon.
Happen. Ears like sails gasping for a wind. You picture
a rope you could hang a man from, you're seeing its tail.

Tusks like bannisters. I almost believe myself. Can you
drum up a roar as wide as a continent, a deep hot note
that bellowed out and belonged to the melting air? You got it.

But people have always lied! You know some say it had a trunk
like a soft telescope, that it looked up along it at the sky
and balanced a bright, gone star on the end, and it died, died.

Carol Ann Duffy

The Swallow

Who is quick, quick, quick,
and lives Lord,
if not I?
Small black arrow
of Your blue sky.
I stun the wind
by the swift ease
of my flight
but, under the eaves of the roof,
in their cosy clay home
my nestlings are hungry.
Quick, quick, quick,
in the hunt for their food,
I dart
from the top to the bottom
of heaven
with a whistle of joy,
then my beak opens
to snap up some inalert fly.
Lord,
a day will come,
a chill gold day
when my babes will take wing
on their own affairs.
Oh! On that day,
when there will be nothing more to bring,
console me
with the call of countries far away.

                             Amen.

Carmen Bernos de Gasztold
Translated by Rumer Godden

The Swan

Silent is my dress when I step across the earth,
reside in my house, or ruffle the waters.
Sometimes my adornments and this high windy air
lift me over the livings of men,
the power of the clouds carries me far
over all people. My white pinions
resound very loudly, ring with a melody,
sing out clearly, when I sleep not on
the soil or settle on grey waters—a travelling spirit.

An Exeter Riddle
Translated by Kevin Crossley-Holland

Swan

Swan, unbelievable bird, a cloud floating,
Arrangement of enormous white chrysanthemums
In a shop kept by angels, feathery statue
Carved from the fall of snow,

You are not too proud to take the crusts I offer.
You are so white that clear water stains you,
And I am ashamed that you have to swim
Here, where cigarette cartons hang in the lake,

And the plastic containers that held our ice-cream.
Now you bend your neck strong as a hawser
And I see your paddles like black rubber
Open and close as you move the webs of your swimming.

About you the small ducks, the coots, and the timid
Water-rails keep their admiring distances. Do not hurry.
Take what you need of my thrown bread, white swan,
Before you drift away, a cloud floating.

Leslie Norris

Hawk Roosting

I sit in the top of the wood, my eyes closed.
Inaction, no falsifying dream
Between my hooked head and hooked feet:
Or in a sleep rehearse perfect kills and eat.

The convenience of the high trees!
The air's buoyancy and the sun's ray
Are of advantage to me;
And the earth's face upward for my inspection.

My feet are locked upon the rough bark.
It took the whole of Creation
To produce my foot, my each feather:
Now I hold Creation in my foot

Or fly up, and revolve it all slowly—
I kill where I please because it is all mine.
There is no sophistry in my body:
My manners are tearing off heads—

The allotment of death.
For the one path of my flight is direct
Through the bones of the living.
No arguments assert my right:

The sun is behind me.
Nothing has changed since I began.
My eye has permitted no change.
I am going to keep things like this.

Ted Hughes

Clock-a-Clay

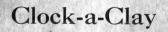

In the cowslip pips I lie,
Hidden from the buzzing fly,
While green grass beneath me lies,
Pearled with dew like fishes' eyes,
Here I lie, a clock-a-clay,
Waiting for the time of day.

While grassy forest quakes surprise,
And the wild wind sobs and sighs,
My gold home rocks as like to fall,
On its pillar green and tall;
When the pattering rain drives by
Clock-a-clay keeps warm and dry.

Day by day and night by night,
All the week I hide from sight;
In the cowslip pips I lie,
In rain and dew still warm and dry;
Day and night, and night and day,
Red, black-spotted clock-a-clay.

My home shakes in wind and showers,
Pale green pillar topped with flowers,
Bending at the wild wind's breath,
Till I touch the grass beneath;
Here I live, lone clock-a-clay,
Watching for the time of day.

John Clare

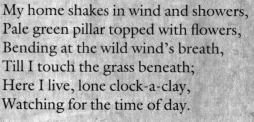

Woodlouse

Armoured dinosaur,
blundering through jungle grass by
dandelion-light.

Knight's headpiece, steel-hinged
orange-segment, ball-bearing,
armadillo-drop.

Pale peppercorn, pearled
eyeball; sentence without end,
my rolling full-stop.

Judith Nicholls

Fleas

Roaming these
Furry prairies,
Daring every so
Often to stop
And sink a well
In the soft pink
Soil, hoping
To draw up a
Hasty drop, and
Drink, and survive,

There's always
The threat of those
Inexplicable storms,
When over the hairy
Horizon rages
A terrible paw:
Descending to
Rend the ground,
While we scramble
Away for our lives.

Valerie Worth

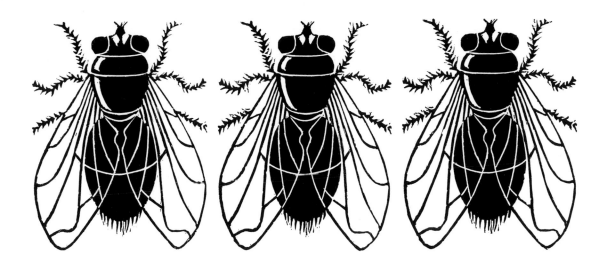

Flies

Flies wear
Their bones
On the outside.

Some show dead
Gray, as bones
Should seem,

But others gleam
Dark blue, or bright
Metal-green,

Or a polished
Copper, mirroring
The sun:

If all bones
Shone so, I
Wouldn't mind

Going around
In my own
Skeleton.

Valerie Worth   99

Chipmunk

Weeping and lamenting,
The chipmunk sits:
'Now where am I going to
Find any nuts!
Haven't I been searching
For days on end?
And didn't I hide them
Under the ground?
Didn't I pack them
Into my cheeks,
Each, my chosen one,
Chosen by me, each!
All kinds of nasty customers
Are on the loose.
Now one's found my storehouse.
Shouldn't he be asking: Whose?
No, he doesn't bother,
Slides his paws right in,
And the nuts are rolling round,
And all he does is grin!
Look, he's stopped, he's smirking,
That old bear's snout!
May you break a tooth
On the hardest one, you lout!'

But a crow makes fun:
'Listen, flabface, lay
A larger stock in next time,
Don't give the game away!'
'Oh I'm so unlucky,
From number one to five,
Stripes along my back,
Mi-se-ra-ble stripes!
My paws aren't worth a penny,
My tail is an excuse!
That ruffian has robbed me,
He's cleaned me out, the brute!
He should have his ears boxed,
Get what he deserves!
Mama, why'd you bring me
Like this into the world,
Such a puny fellow,
Such a nobody.
Listen, Mama chipmunk,
Give birth again to me:
So bitter sorrow shouldn't
Be my lot in life here;
If stripes are what I'm stuck with,
So let me be a tiger!'

Irina Ratushinskaya

The Vixen

Among the taller wood with ivy hung,
The old fox plays and dances round her young.
She snuffs and barks if any passes by
And swings her tail and turns prepared to fly.
The horseman hurries by, shc bolts to see,
And turns agen, from danger never free.
If any stands she runs among the poles
And barks and snaps and drives them in the holes.
The shepherd sees them and the boy goes by
And gets a stick and progs the hole to try.
They get all still and lie in safety sure,
And out again when everything's secure,
And start and snap at blackbirds bouncing by
To fight and catch the great white butterfly.

John Clare

Squirrels

Tails like dandelion clocks
They blow away, these
Light-weight bucking broncos
With a plume behind.

For sheer surprise
No well-aimed burdock
Sticks more nimbly to your overcoat
Than these to tree bark,

Nor with such aplomb
Can any comparable creature
Lead a dance more deftly
Through the branches.

Down to earth again, they
Hold their tums in, little aldermen,
Or sit on tree stumps
Like old ladies knitting socks.

John Mole

102

Harvest Mouse

A sleek, brown acrobat, he climbs
The golden cornstalk till it sways
And sags beneath him. As it swings,
His tail-end twines a neighbour stalk
And balancing with tail and claw
He climbs aloft until he finds
The crisp, ripe, bristly ear of corn;
Then lies along its tilting length,
As if all corn-ears were created
For mice to nibble at . . . and nibbles.

Clive Sansom

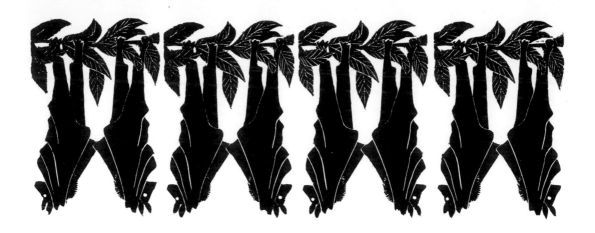

The Bat

The beggarly Bat, a cut out, scattily
Begs at the lamp's light
A bright moth-mote.

What wraps his shivers?
Scraps of moon cloth
Snatched off cold rivers.

Scissored bits
Of the moon's fashion-crazes
Are his disguises
And wrap up his fits—

For the jittery bat's
Determined to burst
Into day, like the sun

But he never gets past
The dawn's black posts.

As long as night lasts
The shuttlecock Bat
Is battered about
By the rackets of ghosts.

Ted Hughes

104

The Bats' Plea

Ignore the stories which say
We shall fly to and tangle your hair,
That you are wise if you dread
Our mouse-like bodies, the way
Our wings fan out and spread
In gloom, in dusty air.

Eagles are lucky to be
Thought of in terms of light
And glory. Half-bird, half-beast,
We're an anomaly.
But, clinging to darkness we rest
And, like stars, belong to the night.

Elizabeth Jennings

The Heron

The Heron stands in water where the swamp
Has deepened to the blackness of a pool,
Or balances with one leg on a hump
Of marsh grass heaped above a muskrat hole.

He walks the shallow with an antic grace.
The great feet break the ridges of the sand,
The long eye notes the minnows' hiding place.
His beak is quicker than a human hand.

He jerks a frog across his bony lip
Then points his heavy bill above the wood.
The wide wings flap but once to lift him up.
A single ripple starts from where he stood.

Theodore Roethke

The Heron

On lonely river-mud a heron alone
Of all things moving—water, reeds and mist—
Maintains his sculptured attitude of stone.
A dead leaf floats on the sliding river, kissed
By its own reflection in a brief farewell.
Movement without sound; the evening drifts
On autumn tides of colour, light, and smell
Of warm decay; and now the heron lifts
Enormous wings in elegy; a grey
Shadow that seems to bear the light away.

Phoebe Hesketh

Something Told The Wild Geese

Something told the wild geese
    It was time to go.
Though the fields lay golden
    Something whispered—'Snow'.
Leaves were green and stirring,
    Berries, lustre-glossed,
But beneath warm feathers
    Something cautioned—'Frost'.

All the sagging orchards
    Steamed with amber spice,
But each wild breast stiffened
    At remembered ice.
Something told the wild geese
    It was time to fly—
Summer sun was on their wings,
    Winter in their cry.

Rachel Field

Complaint of the Wild Goose

We were nine to leave the lake in the north;
Of the nine travellers, I am the last.
Noble hunting falcon, have pity on me.

My country is far; the winds are contrary;
My wings grown heavy, I followed behind, alone.
Noble hunting falcon, have pity on me.

The leaders are at the nest, the stragglers are passing the halting-place;
Winter is coming, the sky is clouded.
Noble hunting falcon, have pity on me.

The sky is grey, the winds rise;
I hear my brothers flying on through the fog.
Noble hunting falcon, have pity on me.

Western Mongolian lullaby
Translated by Willard R. Trask

Lone Dog

I'm a lean dog, a keen dog, a wild dog and lone,
I'm a rough dog, a tough dog, hunting on my own!
I'm a bad dog, a mad dog, teasing silly sheep;
I love to sit and bay at the moon and keep fat souls from sleep.

I'll never be a lap dog, licking dirty feet,
A sleek dog, a meek dog, cringing for my meat.
Not for me the fireside, the well-filled plate,
But shut the door and sharp stone and cuff and kick and hate.

Not for me the other dogs, running by my side,
Some have run a short while, but none of them would bide.
O mine is still the lone trail, the hard trail, the best,
Wide wind and wild stars and the hunger of the quest.

Irene McLeod

The World the First Time

What is that howling, my mother,
Howling out of the sky;
What is it rustles the branches and leaves
And throws the cold snow in my eye?

That is the wind, my wolf son,
The breath of the world passing by,
That flattens the grasses and whips up the lake,
And hurls clouds and birds through the sky.

What is that eye gleaming red, mother,
Gleaming red in the face of the sky;
Why does it stare at me so, mother,
Why does its fire burn my eye?

That is the sun, my wolf child,
That changes dark night into day,
That warms your fur and the pine-needled floor,
And melts the cold snows away.

And who is this serpent that glides, mother,
And winds the dark rocks among,
And laughs and sings as he slides through my paws,
And feels so cold on my tongue?

That is the river, my curious son,
That no creature alive can outrun,
He cuts out the valleys and great watery lakes,
And was here when the world first begun.

And whose is the face that I see, mother,
That face in the water so clear,
Why when I try to catch him
Does he suddenly disappear?

He is closer to you than your brother,
Closer than your father or me,
He'll run beside you your long life through,
For it is yourself that you see.

Gareth Owen

The Cat and the Moon

The cat went here and there
And the moon spun round like a top,
And the nearest kin of the moon,
The creeping cat, looked up.
Black Minnaloushe stared at the moon,
For, wander and wail as he would,
The pure cold light in the sky
Troubled his animal blood.
Minnaloushe runs in the grass
Lifting his delicate feet.
Do you dance, Minnaloushe, do you dance?
When two close kindred meet,
What better than call a dance?
Maybe the moon may learn,
Tired of that courtly fashion,
A new dance turn.
Minnaloushe creeps through the grass
From moonlit place to place,
The sacred moon overhead
Has taken a new phase.
Does Minnaloushe know that his pupils
Will pass from change to change,
And that from round to crescent,
From crescent to round they range?
Minnaloushe creeps through the grass
Alone, important and wise,
And lifts to the changing moon
His changing eyes.

W. B. Yeats

Cats No Less Liquid Than Their Shadows

Cats, no less liquid than their shadows,
Offer no angles to the wind.
They slip, diminished, neat, through loopholes
Less than themselves; will not be pinned

To rules or routes for journeys; counter
Attack with non-resistance; twist
Enticing through the curving fingers
And leave an angered, empty fist.

They wait, obsequious as darkness
Quick to retire, quick to return;
Admit no aim or ethics; flatter
With reservations; will not learn

To answer to their names; are seldom
Truly owned till shot and skinned.
Cats, no less liquid than their shadows,
Offer no angles to the wind.

A. S. J. Tessimond

Horses on the Camargue

In the grey wastes of dread,
The haunt of shattered gulls where nothing moves
But in a shroud of silence like the dead,
I heard a sudden harmony of hooves,
And, turning, saw afar
A hundred snowy horses unconfined,
The silver runaways of Neptune's car
Racing, spray curled, like waves before the wind.
Sons of the Mistral, fleet
As him with whose strong gusts they love to flee,
Who shod the flying thunders on their feet
And plumed them with the snortings of the sea;
Theirs is no earthly breed
Who only haunt the verges of the earth
And only on the sea's salt herbage feed—
Surely the great white breakers gave them birth.
For when for years a slave,
A horse of the Camargue, in alien lands,
Should catch some far-off fragrance of the wave
Carried far inland from his native sands,
Many have told the tale
Of how in fury, foaming at the rein,
He hurls his rider; and with lifted tail,
With coal-red eyes and cataracting mane,

Heading his course for home,
Though sixty foreign leagues before him sweep,
Will never rest until he breathes the foam
And hears the native thunder of the deep.
But when the great gusts rise
And lash their anger on these arid coasts,
When the scared gulls career with mournful cries
And whirl across the waste like driven ghosts:
When hail and fire converge,
The only souls to which they strike no pain
Are the white-crested fillies of the surge
And the white horses of the windy plain.
Then in their strength and pride
The stallions of the wilderness rejoice;
They feel their Master's trident in their side,
And high and shrill they answer to his voice.
With white tails smoking free,
Long streaming manes, and arching necks, they show
Their kinship to their sisters of the sea—
And forward hurl their thunderbolts of snow.
Still out of hardship bred,
Spirits of power and beauty and delight
Have ever on such frugal pastures fed
And loved to course with tempests through the night.

Roy Campbell

The Deer's Request

We are the disappearers.
You may never see us, never,
But if you make your way through a forest
Stepping lightly and gently,
Not plucking or touching or hurting,
You may one day see a shadow
And after the shadow a patch
Of speckled fawn, a glint
Of a horn.
    Those signs mean us.

O chase us never. Don't hurt us.
We who are male carry antlers
Horny, tough, like trees,
But we are terrified creatures,
Are quick to move, are nervous
Of the flutter of birds, of the quietest
Footfall, are frightened of every noise.

If you would learn to be gentle,
To be quiet and happy alone,
Think of our lives in deep forests,
Of those who hunt us and haunt us
And drive us into the ocean.
If you love to play by yourself
Content in that liberty,
Think of us being hunted,
Tell those men to let us be.

Elizabeth Jennings

The Fallow Deer at the Lonely House

One without looks in tonight
    Through the curtain-chink
From the sheet of glistening white;
One without looks in tonight
    As we sit and think
    By the fender-brink.

We do not discern those eyes
    Watching in the snow;
Lit by lamps of rosy dyes
We do not discern those eyes
    Wondering, aglow,
    Fourfooted, tiptoe.

Thomas Hardy

Strippers

If you fall in a river that's full of Piranha,
They'll strip off your flesh like you'd skin a banana.
There's no time for screaming, there's no time for groans.
In forty-five seconds you're nothing but bones.

Dick King-Smith

Rattlesnake Ceremony Song

The king snake said to the rattlesnake:
Do not touch me!
You can do nothing with me.
Lying with your belly full,
Rattlesnake of the rock pile,
Do not touch me!
There is nothing you can do,
You rattlesnake with your belly full,
Lying where the ground-squirrel holes are thick.
Do not touch me!
What can you do to me?
Rattlesnake in the tree clump,
Stretched in the shade,
You can do nothing;
Do not touch me!
Rattlesnake of the plains,
You whose white eye
The sun shines on,
Do not touch me!

Yokuts, North American Indian
Translated by A. L. Kroeber

The Rattlesnake

An ominous length uncoiling and thin,
A sliver of Satan annoyed by the din
Of six berry-pickers, bare-legged and intent
On stripping red treasure like rubies from Ghent.
He moved without motion, he hissed without noise—
A sombre dark ribbon that laughter destroys;
He eyed them unblinking from planets unknown,
As alien as Saturn, immobile as stone.
With almost forbearance, he watched them retreat—
A creature of deserts, and mountains and heat;
No hint of expression, no trace of regret,
No human emotion, a bland baronet
Secure in his duchy, remote and austere,
Ubiquitous, marvellous grass privateer.

Alfred Purdy

Lizard

A lean wizard—
watch me slither
up and down
the breadfruit tree
sometimes pausing a while
for a dither in the sunshine

The only thing
that puts a jitter up my spine
is when I think about
my great great great
great great great great
great great grandmother
Dinosaura Diplodocus

She would have the shock of her life
if she were to come back
and see me reduced to lizardsize!

Grace Nichols

Iguana Memory

Saw an iguana once
when I was very small
in our backdam backyard
came rustling across my path

green like moving newleaf sunlight

big like big big lizard
with more legs than centipede
so it seemed to me
and it must have stopped a while
eyes meeting mine
iguana and child locked in a brief
split moment happening
before it went hurrying

for the green of its life

Grace Nichols

123

The Sloth

In moving-slow he has no Peer.
You ask him something in his ear;
He thinks about it for a Year;

And, then, before he says a Word
There, upside down (unlike a Bird)
He will assume that you have Heard—

A most Ex-as-per-at-ing Lug.
But should you call his manner Smug,
He'll sigh and give his Branch a Hug;

Then off again to Sleep he goes,
Still swaying gently by his Toes,
And you just *know* he knows he knows.

Theodore Roethke

The Jaguar

The jaguar lies secretly in trees,
Whence with abandon and gymnastic ease
It drops upon you as you walk below,
And breaks your neck with one effective blow.

The Amazonian, on his native plot,
Refers to them as 'tigers', which they're not.
The difference, as you expire like this,
Must be however pure hypothesis.

Dick King-Smith

Don't Call Alligator Long-Mouth Till You Cross River

Call alligator long-mouth
call alligator saw-mouth
call alligator pushy-mouth
call alligator scissors-mouth
call alligator raggedy-mouth
call alligator bumpy-bum
call alligator all dem rude word
but better wait
                    till you cross river.

John Agard

Alligator

If you want to see an alligator
you must go down to the muddy slushy end
of the old Caroony River

I know an alligator
who's living down there
She's a-big. She's a-mean. She's a-wild.
She's a-fierce.

But if you really want to see an alligator
you must go down to the muddy slushy end
of the old Caroony River

Go down gently to that river and say
'Alligator Mama
Alligator Mama
Alligator Mamaaaaaaaa'

And up she'll rise
but don't stick around
RUN FOR YOUR LIFE

Grace Nichols

The Eagle Above Us

In the sky the eagle
There is his place, there far above us.
Now he appears there
He holds the world fast in his talons.
The world has put on a grey dress
A beautiful, living, watery dress of clouds.
There he is, far above us
In the midst of the sky.
Shining, he looks down on his world.
His countenance is full of terrible disaster.
His eye is glorious.
His feet are already dark red.
There he is far above us in the middle of the sky.
There he remembers those who have been on earth.
He spreads his wings over them.

North American Indian
Translated by Willard R. Trask

Humming Bird

The humming bird refuels
in mid-air from the hub
of a fuchsia flower.
Its belly is feathered white
as rapids; its eye
is smaller than a drop of tar.
A bodied moth, it beats
stopwatches into lethargy
with its wingstrokes.

Food it needs every fifteen
minutes. It has the metabolism
of a steam engine.
Its tiny claws are slight
as pared fingernail;
you could slip it with ease
into a breast pocket.
There it might lie, cowed
—or give you a second heart.

Paul Groves

The Condor

I am the biggest bird there ever was.
My wings are like barn doors.
I come from the plains, by the mountains bare,
By the moorland grasses.
I am first on the scene at an accident.
My bill is a gravedigger's spade.
I circle your dying body.
I pick your bones white.
I leave you bleached in the wilderness.
I drop on your carcass
Like the angel of death.
I am the giant condor
My black wings span like gates to the land of the dead.

David Liptrot

130

The Cat-eyed Owl

The cat-eyed owl, although so fierce
At night with kittens and with mice

In daylight may be mobbed
By flocks of little birds, and in
The market-place, be robbed

Of all his dignity and wisdom
By children market-women and malingering men

Who hoot at it and mocking its myopic
Eyes, shout: 'Look!
Look at it now, he hangs his head in
Shame.' This never happens to the eagle
Or the nightingale.

Edward Kamau Brathwaite

Pumas

Hushed, cruel, amber-eyed,
Before the time of the danger of the day,
Or at dusk on the boulder-broken mountainside
The great cats seek their prey.

Soft-padded, heavy-limbed,
With agate talons chiselled for love or hate,
In desolate places wooded or granite-rimmed,
The great cats seek their mate.

Rippling, as water swerved,
To tangled coverts overshadowed and deep
Or secret caves where the canyon's wall is curved,
The great cats go for sleep.

Seeking the mate or prey,
Out of the darkness glow the insatiate eyes.
Man, who is made more terrible far than they,
Dreams he is otherwise!

George Sterling

Puma

Within the Puma's golden head
   burn Jungle, flames and paradises.
And all who look into his red
and fiery eye by fury fed
   the Puma paralyses.

The Sun, the Orchid and the Snake
   like demons of the Bible
rage in his brow. The earthquake
and the volcanic mountain shake
   and tremble in his eyeball.

George Barker

The Chipmunk's Day

In and out the bushes, up the ivy,
Into the hole
By the old oak stump, the chipmunk flashes.
Up the pole

To the feeder full of seeds he dashes,
Stuffs his cheeks,
The chickadee and titmouse scold him.
Down he streaks.

Red as the leaves the wind blows off the maple,
Red as a fox,
Striped like a skunk, the chipmunk whistles
Past the love seat, past the mailbox,

Down the path,
Home to his warm hole stuffed with sweet
Things to eat.
Neat and slight and shining, his front feet

Curled at his breast, he sits there while the sun
Stripes the red west
With its last light: the chipmunk
Dives to his rest.

Randall Jarrell

Skunk

Sometimes, around
Moonrise, a wraith
Drifts in through
The open window:
A vague cold taint
Of rank weeds
And phosphorescent
Mold, a hint
Of obscure dank
Root hollows and
Mist-woven paths,
Pale toadstools and
Dark-reveling worms:
As the skunk walks
By, half vapor, half
Shade, diffusing
The night's uncanny
Essence and atmosphere.

Valerie Worth

Coyote, or The Prairie Wolf

Blown out of the prairie in twilight and dew,
Half bold and half timid, yet lazy all through,
Loth ever to leave, and yet fearful to stay,
He limps in the clearing, —an outcast in grey.

A shade on the stubble, a ghost by the wall,
Now leaping, now limping, now risking a fall,
Lop-eared and large jointed, but ever alway
A thoroughly vagabond outcast in grey.

Here, Carlo, old fellow, he's one of your kind, —
Go seek him, and bring him in out of the wind.
What! snarling, my Carlo! So—even dogs may
Deny their own kin in the outcast in grey!

Well, take what you will, —though it be on the sly,
Marauding or begging, —I shall not ask why;
But will call it a dole, just to help on his way
A four-footed friar in orders of grey!

Bret Harte

I Rise, I Rise

I rise, I rise,
I, whose tread makes the earth to rumble.
I rise, I rise,
I, in whose thighs there is strength.
I rise, I rise,
I, who whips his back with his tail when in rage.
I rise, I rise,
I, in whose humped shoulder there is power.
I rise, I rise,
I, who shakes his mane when angered.
I rise, I rise,
I, whose horns are sharp and curved.

Osage, North American Indian
From a prayer before a young man's
first buffalo hunt

Buffalo Dusk

The buffaloes are gone.
And those who saw the buffaloes are gone.
Those who saw the buffaloes by thousands and
    how they pawed the prairie sod into dust
    with their hoofs, their great heads down
    pawing on in a great pageant of dusk,
Those who saw the buffaloes are gone.
And the buffaloes are gone.

Carl Sandburg

The Flower-fed Buffaloes

The flower-fed buffaloes of the spring
In the days of long ago,
Ranged where the locomotives sing
And the prairie flowers lie low: —
The tossing, blooming, perfumed grass
Is swept away by the wheat,
Wheels and wheels and wheels spin by
In the spring that still is sweet.
But the flower-fed buffaloes of the spring
Left us, long ago.
They gore no more, they bellow no more
They trundle around the hills no more: —
With the Blackfeet, lying low,
With the Pawnees, lying low,
Lying low.

Vachel Lindsay

The Grizzly Bear

I see a bear
Growing out of a bulb in wet soil licks its black tip
With a pink tongue its little eyes
Open and see a present an enormous bulging mystery package
Over which it walks sniffing at seams
Digging at the wrapping overjoyed holding the joy off sniffing
    and scratching
Teasing itself with scrapings and lickings and the thought of it
And little sips of the ecstasy of it

O bear do not open your package
Sit on your backside and sunburn your belly
It is all there it has actually arrived
No matter how long you dawdle it cannot get away
Shamble about lazily laze there in the admiration of it
With all the insects it's attracted all going crazy
And those others the squirrel with its pop-eyed amazement
The deer with its pop-eyed incredulity
The weasel pop-eyed with envy and trickery
All going mad for a share wave them off laze
Yawn and grin let your heart thump happily
Warm your shining cheek fur in the morning sun

You have got it everything for nothing

Ted Hughes

Song of a Bear

There is danger where I move my feet.
I am a whirlwind. There is danger where I move my feet.
I am a gray bear.
When I walk, where I step lightning flies from me.
Where I walk, one to be feared.
Where I walk, long life.
One to be feared I am.
There is danger where I walk.

Navajo, North American Indian

139

Mooses

The goofy Moose, the walking house-frame,
Is lost
In the forest. He bumps, he blunders, he stands.

With massy bony thoughts sticking out near his ears—
Reaching out palm upwards, to catch whatever might be falling from heaven—
He tries to think,
Leaning their huge weight
On the lectern of his front legs.

He can't find the world!
Where did it go? What does a world look like?
The Moose
Crashes on, and crashes into a lake, and stares at the mountain, and cries
'Where do I belong? This is no place!'

He turns and drags half the lake out after him
And charges the cackling underbrush—

He meets another Moose.
He stares, he thinks 'It's only a mirror!'

'Where is the world?' he groans, 'O my lost world!
And why am I so ugly?
And why am I so far away from my feet?'

He weeps.
Hopeless drops drip from his droopy lips.

The other Moose just stands there doing the same.

Two dopes of the deep woods.

Ted Hughes

Those Game Animals

Those game animals, those long-haired caribou,
Though they roam everywhere, I am quite unable to get any.
I carried this bow of mine in my hand always.
At last I pondered deeply:
It is all right, even if
I am quite unable to get them in the present winter.

Those game animals, those seals,
Though they keep visiting their holes, I am quite unable to get any.
I carried this harpoon of mine in my hand always.
At last I pondered deeply:
It is all right, even if
I begin at last to be greatly afraid in this present summer.

Those game animals, those fish.
Though I go out in the middle of the lake, I am quite unable to get any.
At last I pondered deeply:
It is all right, even if
I begin at last to be afraid of the hummocky ice within.

Those seals, those fearful brown bears,
Constantly walking about here, I begin to be terrified.
This arrow of mine is fearless, this arrow.
Am I to allow myself to be terrified at last?

Innuit, Canada

Seaside Suicide

When Lemmings migrate, and arrive at a river,
They plummet right in it with never a shiver
And swim till their feet touch the ground.

When Lemmings migrate, and arrive at a lake,
They swim straight across it, and out with a shake,
And onward relentlessly bound.

And then, when the Lemmings' migratory host
Has crossed the whole land and arrived at the coast,
They swim out to sea and are drowned.

Dick King-Smith

The Lemmings

Once in a hundred years the Lemmings come
Westward, in search of food, over the snow;
Westward until the salt sea drowns them dumb;
Westward, till all are drowned, those Lemmings go.

Once, it is thought, there was a westward land
(Now drowned) where there was food for those starved things,
And memory of the place has burnt its brand
In the little brains of all the Lemming kings.

Perhaps, long since, there was a land beyond
Westward from death, some city, some calm place
Where one could taste God's quiet and be fond
With the little beauty of a human face;

But now the land is drowned. Yet still we press
Westward, in search, to death, to nothingness.

John Masefield

Polar Bear

I saw a polar bear
on an ice-drift.
He seemed harmless as a dog,
who comes running towards you,
wagging his tail.
But so much
did he want to get at me
that when I jumped aside
he went spinning on the ice.
We played this game of tag
from morning until dusk.
But then at last, I tired him out,
and ran my spear into his side.

Iglukik
Translated by T. Lowenstein

I Saw Two Bears

I saw two Bears, as white as any milk,
Lying together in a mighty cave,
Of mild aspect, and hair as soft as silk,
That savage nature seemed not to have,
Nor after greedy spoil of blood to crave;
Two fairer beasts might not elsewhere be found,
Although the compassed world were sought around.
But what can long abide above this ground
In state of bliss, or steadfast happiness?

Edmund Spenser

The Arctic Fox

No feet. Snow.
Ear—a star-cut
Crystal of silence.
The world hangs watched.

Jaws flimsy as ice
Champ at the hoar-frost
Of something tasteless—
A snowflake of feathers.

The forest sighs.
A fur of breath
Empty as moonlight
Has a blue shadow.

A dream twitches
The sleeping face
Of the snow-lit land.

When day wakes
Sun will not find
What night hardly noticed.

Ted Hughes

Amulet

Inside the Wolf's fang, the mountain of heather.
Inside the mountain of heather, the Wolf's fur.
Inside the Wolf's fur, the ragged forest.
Inside the ragged forest, the Wolf's foot.
Inside the Wolf's foot, the stony horizon.
Inside the stony horizon, the Wolf's tongue.
Inside the Wolf's tongue, the Doe's tears.
Inside the Doc's tears, the frozen swamp.
Inside the frozen swamp, the Wolf's blood.
Inside the Wolf's blood, the snow wind.
Inside the snow wind, the Wolf's eye.
Inside the Wolf's eye, the North Star.
Inside the North Star, the Wolf's fang.

Ted Hughes

In Daylight Strange

It was last Friday at ten to four I
Thought of the lion walking into the playground.
I was sitting, thinking, at our table when
The thought of the lion simply came,
And the sun was very hot, and the lion
Was in the yard (in daylight strange, because
Lions go out at night). He was
An enormous, sudden lion and he
Appeared just like that and was crossing very
Slowly the dusty playground, looking
To neither side, coming towards the door. He was
Coloured a yellow that was nearly grey, or a
Grey that was nearly yellow. He was so
Quiet that only I could hear the huge feet
Solidly pacing, and at the playground door he
Stopped, and looked powerfully in. There was
A forest following him, out in the street,
And noises of parakeets. When he stopped,
Looking like a picture of a lion in the frame

Of the open door, his eyes looked on at
Everything inside with a stern, curious look, he
Didn't seem completely to understand. So
He waited a second or two before
He roared. All the reeds on the river bank
Trembled, a thousand feet
Scattered among the trees, birds rose in clouds
But no one jumped in the classroom, no one screamed,
No one ran to ring the firebell, and
Miss Wolfenden went on writing on the board.
It was just exactly as if
They hadn't heard at all, as if nobody had heard.
And yet I had heard, certainly.
Yes. I had heard,
And I didn't jump.
And would you say you were surprised? Because
You ought not to be surprised.
Why should I be frightened when it was
Because *I* thought of the lion, that the lion was there?

Alan Brownjohn

Index of Animals

Index of Artists

The illustrations are by:

Hamish Blakely
    pp 48–49, 86–87, 94–95, 132–133

Camilla Charnock
    pp 21, 26–27, 42–43, 98–99, 104–105

Jacky Corner
    pp 34–35, 102–103, 108–109, 116–117, 144–145

Zoë Hancox
    pp 24–25, 56–57, 84–85, 139, 142–143

David Holmes
    pp 75, 92–93, 130–131

Jo Lamb
    pp 22–23, 40–41, 120–121, 124–125

Mick Manning
    pp 30–31, 101, 106–107

Alan Marks
    pp 47, 54, 70–71, 79, 114–115, 136–137, 146

Diana Mayo
    pp 64–65, 68–69

Jackie Morris
    endpapers, half title, contents and pp 14–15, 29, 45, 50–51, 61, 91, 113, 119, 148–149

Jane Ormes
    pp 16–17, 32–33, 76–77, 80–81, 96–97, 129

The Rooftop Art Company
    pp 38–39, 59, 72, 88–89, 122–123, 126–127

The jacket illustration is by Jackie Morris

Index of Authors

Index of Titles and First Lines

153

154

Acknowledgements

The editors and publisher are grateful for permission to include the following copyright material:

Aboriginal Aranda: 'Ringneck Parrots', translated by T. G. H. Strehlow, reprinted in *The New Oxford Book of Australian Verse*. **John Agard**: 'Don't Call Alligator Long-Mouth Till You Cross River' from *Say It Again, Granny* (Bodley Head). Reprinted by permission of Random Century Ltd. **Zoë Bailey**: 'Dolphins'. Reprinted by permission of the author. **George Barker**: 'Ape' and 'Puma' from *Alphabetical Zoo*. Reprinted by permission of Faber & Faber Ltd.

Noëline Barry: 'The Red-winged Lourie' from *Mambo Book of Zimbabwean Verse in English*, ed. Colin Style. Reprinted by permission of Mambo Press, Zimbabwe. **Gerard Benson**: 'Bei-shung' from *Headlines from the Jungle* (Penguin). Reprinted by permission of the author. **Edward Kamau Brathwaite**: 'The Cat-eyed Owl', from *Black Poetry* (Blackie, 1988), © Edward Kamau Brathwaite 1988. **Alan Brownjohn**: 'Skate', 'Ostrich', 'Elephant' and 'In Daylight Strange'. All reprinted by permission of the author. **Roy Campbell**: 'The Theology of Bongwi, the Baboon' and 'Horses on the Camargue'.

Reprinted by permission of Francisco Campbell Custodio and Ad. Donker (Pty) Ltd. **John Cassidy**: 'Sea Lions off Monterey' from *Walking on Frogs* (Bloodaxe Books Ltd.). **Ann Coleridge**: 'Possums', © Ann Coleridge. **Kevin Crossley-Holland**: 'The Swan' from *The Exeter Book of Riddles* (Penguin). Reprinted by permission of Rogers Coleridge & White Ltd., Literary Agency. **Carol Ann Duffy**: 'The Legend' from *The Orange Dove of Fiji*, edited by Simon Rae (Hutchinson). Reprinted by permission of Random Century Ltd. on behalf of the World Wide Fund for Nature. **Richard Eberhart**: 'Sea-hawk' from *Collected Poems* (Chatto), © Richard Eberhart. **Charles Eglington**: 'Cheetah' from *New Book of South African Verse* (Oxford University Press, Cape Town, 1979). **Gavin Ewart**: 'Gondwanaland', 'The World Is Full Of Elephants' and 'The Meerkats of Africa' from *Caterpillar Stew* (Hutchinson) and 'Zebra' from *Learned Hippopotamus* (Hutchinson). Reprinted by permission of Random Century Ltd. on behalf of the author. **Eleanor Farjeon**: 'Flamingo' from *Silver, Sand and Snow* (Michael Joseph). Reprinted by permission of David Higham Associates Ltd. **Harold Farmer**: 'Rhinoceros' from *Mambo Book of Zimbabwean Verse In English*, ed. Colin Style. Reprinted by permission of Mambo Press, Zimbabwe. **Rachel Field**: 'Something Told The Wild Geese' from *Poems*. © 1934 by Macmillan Publishing Co., renewed 1962 by Arthur S. Pederson. Reprinted by permission of Macmillan Publishing Co. **Fulani**: 'Chain-Song' from *African Poetry For Schools I* (Longman). **Roy Fuller**: 'The Giraffes' from *New and Collected Poems*. Reprinted by permission of Martin Secker & Warburg Ltd. **Carmen Bernos de Gasztold**: translated by Rumer Godden, 'The Camel', 'The Gazelle' and 'The Swallow' from *The Beasts' Choir* (Macmillan London Ltd.). **Broughton Gingell**: 'Scorpions Fighting' from *Mambo Book of Zimbabwean Verse In English*, ed. Colin Style. Reprinted by permission of Mambo Press, Zimbabwe. **Irene Gough**: 'Under the Range', © Irene Gough. **Jonathan Griffin**: 'Dolphins' from *Headlines From The Jungle*. Reprinted by permission of Kathleen Griffin. **William Hart-Smith**: 'Galahs' from *The Talking Clothes* (Angus & Robertson), © William Hart-Smith. **Phoebe Hesketh**: 'The Heron' from *Prayer for the Sun* (Bodley Head). **Bruce Hewett**: 'The Crocodile' from *The New Book of South African Verse* (Oxford University Press, Cape Town, 1979), © Bruce Hewett 1979. **Hindi**: 'The Shot Deer' from *Distant Voices*, ed. Denys Thompson. Used by permission of Mrs M. E. Thompson. **Hsu Pen**: 'Ballad of the Ferocious Tiger', from *The Columbia Book of Later Chinese Poetry*. Reprinted by permission of Columbia University Press. **Ted Hughes**: 'The Grizzly Bear', 'Mooses', 'The Arctic Fox' and 'Amulet' from *Under the North Star*. Reprinted by permission of Faber & Faber Ltd. [Viking]. 'The Bat' from *What is the Truth*. Reprinted by permission of Faber & Faber Ltd. [Harper]. 'Hawk Roosting' from *The Hawk in the Rain*. Reprinted by permission of Faber & Faber Ltd. [Harper]. **Igbo**: 'You!' first published in *Poetic-Heritage* (Nwanko-Ifejika & Co., Nigeria, 1971). **Iglukik**: translated by Tom Lowenstein, 'Polar Bear' from *Eskimo Poems* (Allison & Busby). Translation © Tom Lowenstein. **Innuit**: 'Those Game Animals' from *Distant Voices*, ed. Denys Thompson. Used by permission of Mrs M. E. Thompson. **Randall Jarrell**: 'The Chipmunk's Day' from *The Bat Poet*. **Elizabeth Jennings**: 'The Bats' Plea' and 'The Deer's Request' from *After the Ark* (Oxford University Press). Reprinted by permission of David Higham Associates Ltd. **Jenifer Kelly**: 'If You Go Softly', © Jenifer Kelly. **Khoikhoi**: 'Song of the Lioness for her Cub', translated by Ulli Beier from *African Poetry* (Cambridge University Press, 1966). **Dick King-Smith**: 'The Jaguar' from *Alphabeasts*. Reprinted by permission of Victor Gollancz Ltd. 'Strippers' and 'Seaside Suicide' from *Jungle Jingles* (Doubleday), © 1990 by Dick King-Smith. Reprinted by permission of Transworld Publishers Ltd. **Jejuri Arun Kolatkar**: 'The Butterfly' from *I Like That Stuff* (Cambridge University Press). © Jejuri Arun Kolatkar. **D. H. Lawrence**: 'Kangaroo' from *The Complete Poems of D. H. Lawrence*, © 1964, 1971 by Angelo Ravagli and C. M. Weekley, executors of the Estate of Frieda Lawrence Ravagli. Used by permission of Viking Penguin, a division of Penguin Books USA Inc. **Vachel Lindsay**: 'The Flower-fed Buffaloes' from *Going To The Stars*. © 1926 by D. Appleton & Co., renewed 1954 by Elizabeth C. Lindsay. Reprinted by permission of Penguin USA. **Li Po**: 'Autumn Cove' from *Chinese Lyricism*. Reprinted by permission of Columbia University Press. **David Liptrot**: 'The Condor', © 1992 David Liptrot. Reprinted by permission of the author. **Douglas Livingstone**: 'Vulture' from *Sjambok and Other Poems from Africa*, © OUP 1964. Reprinted by permission of Oxford University Press. **John Mbiti**: 'The Snake Song'. Reprinted by permission of the author. **Irene McLeod**: 'Lone Dog' from *Songs to Save a Soul* (Chatto & Windus). Reprinted by permission of Random Century Ltd. on behalf of the Estate of Irene McLeod. **Rupendra Guha Majumdar**: 'In the Garden' from *Indo-English Poetry in Bengal*, ed. K. C. Lahivi, Writers Workshop, Calcutta, 1974. **John Masefield**: 'The Lemmings' from *Collected Poems* (Heinemann). Reprinted by permission of The Society of Authors as the literary representative of the Estate of John Masefield. **John Mole**: 'Squirrels' from *Boo To A Goose* (Peterloo Poets, 1987). Reprinted with permission. **Edwin**

156

Morgan: 'Hyena' from *From Glasgow to Saturn*. Reprinted by permission of Carcanet Press. **Navajo**: 'Song of a Bear' from *The Unwritten Song* Vol. 2. (Macmillan London Ltd.). **Judith Nicholls**: 'Whalesong' and 'Orang-utan' from *Dragonsfire*, © Judith Nicholls 1990. 'Woodlouse' from *Midnight Forest*, © Judith Nicholls 1987. Reprinted by permission of Faber & Faber Ltd. **Grace Nichols**: 'Lizard' and 'Alligator' from *Come Into My Tropical Garden*, © Grace Nichols. 'Iguana Memory' from *Fat Black Woman's Poems*. Reprinted by permission of Virago Press. **Leslie Norris**: 'Swan', © Leslie Norris. **A. K. Nyabongo** (translator): 'Mother Parrot's Advice to her Children' from *African Poetry for Schools I* (Longman). **Osage**: 'I Rise, I Rise' from *Dancing Teepees*, ed. Virginia Driving Hawk Sneve (1990). From an Osage prayer, first published in Thirty-ninth Annual Report of the Bureau of American Ethnology 1917–1918 (Smithsonian Institute, Washington, 1925). **Gareth Owen**: 'The World the First Time'. Used by permission of the author. **Neil Paech**: 'Parrots', © Neil Paech. **Lydia Pender**: 'Flying Foxes', © Lydia Pender. **E. J. Pratt**: 'The Shark', © Viola & Claire Pratt and the Pratt Estate. **Alfred Purdy**: 'The Rattlesnake' from *The New Wind Has Wings* (Oxford University Press, Toronto 1984). Used by permission of the author. **Irina Ratushinskaya**: 'Chipmunk' from *Pencil Letter* (Bloodaxe Books Ltd.). **Tom Rawling**: 'The Names of the Sea-trout' from *The Old Showfield*. © Tom Rawling. **Michael Richards**: 'Penguin', © 1992 Michael Richards. Used with permission. The extract from *Life On Earth* by David Attenborough is reproduced by permission of HarperCollins Publishers. **Theodore Roethke**: 'The Heron' and 'The Sloth' from *Collected Poems*. Reprinted by permission of Faber & Faber Ltd. [Doubleday]. **Carl Sandburg**: 'Buffalo Dusk' from *Smoke and Steel* (Harcourt Brace Jovanovich). **Clive Sansom**: 'Harvest Mouse' from *An English Year* (Chatto). Reprinted by permission of David Higham Associates Ltd. **Shona** (Traditional): 'The Praises of Baboon, the Totem Animal' translated by George Fortune, from *Mambo Book of Zimbabwean Verse in English*. Reprinted by permission of Mambo Press, Zimbabwe. **Iain Crichton Smith**: extract from 'Australia'. Reprinted by permission of Carcanet Press. **Sotho**:

'Leopard' from *African Poetry for Schools I* (Longman). **Anthony Stuart**: 'The Legend of the Panda', © 1992 Anthony Stuart. Used with permission. **Colin Style**: 'Herd of Impala' from *Mambo Book of Zimbabwean Verse in English*, ed. Colin Style. Reprinted by permission of Mambo Press, Zimbabwe. **A. S. J. Tessimond**: 'Cats No Less Liquid Than Their Shadows' from *The Walls of Glass* (Methuen). Reprinted by permission of Hubert Nicholson. **Margaret Toms**: 'Hippo' from *The Unicorn and the Lions* (Macmillan London Ltd.). **Madagascan**: 'The Locust', translated by A. Marre and Willard R. Trask, from *Junior Voices 3*. **Willard R. Trask**: 'The Complaint of the Wild Goose' and 'The Eagle Above Us' from *The Unwritten Song*, Vol. 2, ed. with translations by Willard R. Trask. © 1967 by Willard R. Trask. Reprinted by permission of Macmillan Publishing Co. **Chris Wallace-Crabbe**: 'A Glimpse of Shere Khan' from *I'm Deadly Serious*, © Chris Wallace-Crabbe 1988. Reprinted by permission of Oxford University Press. **Marilyn Watts**: 'On the Serengeti', © 1992 Marilyn Watts. Used with permission. **Valerie Worth**: 'Octopus', 'Fleas', 'Flies' and 'Skunk' from *Small Poems Again*, © 1975, 1986 by Valerie Worth. Reprinted by permission of Farrar, Straus and Giroux, Inc. **Judith Wright**: 'Egrets' and 'Night Herons' from *Collected Poems* (Angus & Robertson), © Judith Wright. **Kit Wright**: 'The Song of the Whale' from *Hot Dog and Other Poems* (Kestrel Books, 1981), © Kit Wright 1981. Reprinted by permission of Penguin Books Ltd. **W. B. Yeats**: 'The Cat and the Moon' from *Collected Poems 1*, © 1919 by Macmillan Publishing Co., renewed 1947 by Bertha Georgie Yeats. Reprinted by permission of Macmillan Publishing Co. **Yoruba** (Anon.): translated by Ulli Beier, 'Kob Antelope', 'Buffalo' and 'Hyena' from *The Rattle Bag* (Faber & Faber Ltd).

While every effort has been made to trace and contact copyright holders, this has not always been possible. If contacted, the publisher will be pleased to correct any errors or omissions at the earliest opportunity.

The editors and publisher wish to thank the Department of Zoology at Oxford University, for help and information.